NEW ZEALAND WARBIRDS

NEW ZEALAND WARBIRDS

JOHN KING

CENTURY HUTCHINSON

To Dorothy

Century Hutchinson New Zealand Ltd
An imprint of the Century Hutchinson Group
187 Archers Road, P.O. Box 40-086, Glenfield, Auckland 10.

Century Hutchinson Ltd
62-65 Chandos Place, Covent Garden, London WC2N4NW

Century Hutchinson Australia Pty Ltd
20 Alfred Street, Milsons Point, N.S.W. 2061

Century Hutchinson South Africa Pty Ltd
P.O. Box 337, Bergvlei 2012, South Africa

First Published 1989
©John King 1989

Printed in Hong Kong

ISBN 1 86941 036 X

CONTENTS

FOREWORD

The New Zealand Warbirds Association was born out of an idea generated by a small group of enthusiasts who lamented the sale of New Zealand's only flyable P-51D Mustang, ZK-CCG, to overseas interests. With the sale of the RNZAF's Harvard trainers imminent, we were determined to salvage as many as possible so they would continue to be seen in New Zealand skies for many years to come.

From those small beginnings back in 1978, with Gerard Brown as secretary and Ernie Thompson and Brian Rhodes providing moral support, the New Zealand Warbirds Association has grown to become the largest organisation of its kind in the Southern Hemisphere. Warbirds members own and operate a large collection of vintage and historic aircraft and have become a major drawcard at aviation displays throughout the country.

The growth of New Zealand Warbirds has been spectacular, and has been made possible only by the abundance of enthusiasm and goodwill generated by the membership as a whole. Contrary to some opinion, membership of the association has never been restricted and is open to anyone who shares the same enthusiasm to keep classic and vintage aircraft flying.

Throughout the 10 years of the association's existence we have enjoyed a close working relationship with the RNZAF Museum Trust and the Flight Operations Division, Civil Aviation Division, Ministry of Transport, without whose assistance we could not have matured into the organisation that we are today.

The future looks very exciting, with the arrival of Tim Wallis's Supermarine Spitfire MK.XVI, and the P-40 Kittyhawk due to fly within the next 12 months. The ultimate goal of New Zealand Warbirds is to have a first-class facility at Ardmore to house all the members' aircraft. This complex would be open to the public daily, and regular 'live weekends' would be scheduled throughout the year.

New Zealand Warbirds thrives on mutual co-operation among its members, and through their enthusiasm and support I believe our aims will be achieved.

T. T. BLAND
President

Ardmore-based Harvards in front of Auckland's symbol, Rangitoto Island.

THE PILOTS

Although membership is open to everybody and a wide range of people display aeroplanes that are very different, Warbirds pilots are special.

We have quite a lot in common. Because our speciality is airshows we have to be good at them. This entails all our flyers co-operating in an outstanding manner to achieve the sort of result which might otherwise be expected only from air force pilots who have been flying together for years.

The launch of the America's Cup challenger *KZ1* in April 1988 was a good example. Twenty-five aircraft carried out displays within a total time of 10 minutes, and the display started within 10 seconds and finished within 20 seconds of the planned time with no errors or infringements.

All credit is due to the pilots. They knew it had to be right so they made it right. Flight discipline is exacting in such cases. These pilots listened to the briefing, flew precisely and had the common sense and wisdom to know where they could ad lib to make it work, for example when the DC.3 would not start or when similar problems have occurred.

Our display pilots are left in no doubt where they stand as to what they may or may not do. If they want to do more, we train them. If they are not sure, we check them.

To work as closely as we do it is necessary to know and trust the other people implicitly. We help each other with pleasure because we enjoy working together. From this comes *esprit de corps,* and from this comes the results seen by everybody else. We work together before and after the glamour bits, and those who put the most into it usually seem to get the most out of it. We actively encourage those who are keen and show ability to get the most out of their flying by obtaining display clearances, and as time goes by we will have a considerable pool of talent to draw from.

Not all can or want to achieve these things. A small and competent team works behind the scenes and is appreciated by those who know who they are. There are now about 400 members in the Warbirds Association and still nearly all the work is done by the few, so any new members who feel left out of what is going on should realise it is not intentional. New members are recommended to grab hold with both hands and take part.

Bill West in Harvard NZ1065.

We do our best to look after our reputation, but because of the high profile of these types of aircraft we are in the public eye, which seems to make some people feel entitled to criticise the pilots or the operation at times. This tends to leave us rather bemused because we feel that, overall, we are doing very well.

The recent addition of the Airtrainer to the fleet will, I hope, lead to a further crop of pilots becoming skilled enthusiasts who enjoy their flying, and along the way provide much pleasure for others.

JOHN DENTON
Chief Flying Instructor 1978–1987

This paint scheme was used by the RNZAF in the late 1960s, the extra engine cowling adornment being for the Red Checkers aerobatic team. Strictly speaking the roundel should have a silver fern in the centre, but one of the owners of NZ1098 doesn't like the thought of 'white feathers'.

HARVARD

Overall training yellow was a typical wartime RNZAF Harvard colour scheme.

Big, blunt and noisily spectacular, the North American AT-6 Harvard has a very special meaning for the New Zealand Warbirds Association, for it was around this type of aircraft that the movement was founded.

The international warbirds groups tended to concentrate on surplus fighters and bombers at the top end of the scale, with the lesser trainers making up the numbers. In New Zealand the surplus RNZAF fighter stocks of Kittyhawks, Corsairs and Mustangs had been sold, mostly for scrap, by the time enough interest had been shown in such aircraft to suggest that they might be preserved, in flying condition, for future generations to marvel at. The future did not look particularly bright.

But things had changed by 1977, when the RNZAF gathered together all 19 Harvard trainers left flying (out of a total of 202 brought into the country since March 1941), plus all those in storage and in use as instructional airframes around the country, and put them up for tender. The opportunity was unique in New Zealand, and it was feared that all the Harvards would be sold to overseas buyers. Happily this was not to be. Certainly some did go to America, but the main bidder wanted only the Pratt & Whitney R-1340 engines and most of the airframes stayed in New Zealand after some further trading. Others were flown to Australia, where they sat grounded for a long time while the Department of Transport considered whether it should allow them to continue flying.

The first private owners of these beefy military trainers were mostly ex-RNZAF pilots who remembered them from their early training days, but who had gone into airline flying and could afford not only the initial price but also the cost of the large amount of fuel needed to keep them flying. A few others were aero-club-trained pilots who had always hankered after something with a bit of grunt to it, and syndicates were formed, the floating membership of which has always made it difficult to discover just exactly who owns which Harvard.

NZ1025, ZK-ENN has been easy to keep track of, however. It was bought by Dave Diamond and Jim Sullivan, a pair of South Canterbury farmers who put it into the air as the first to be granted a civilian certificate of airworthiness and it was based for a decade at Timaru. More recently purchased by John Greenstreet, it now shares the main Warbirds hangar at Ardmore with the rest of the Harvard squadron.

The North American Aviation company was born in 1934 out of a series of company mergers, and under James 'Dutch' Kindelberger went after a US Army Air Corps contract for a primary trainer as its first major design project. The resulting NA-16 was chosen by the military, accepted and formed the basis of a variety of trainers and light attack bombers, with engines ranging from the 400 hp Wright R-975 to the Pratt & Whitney Wasp R-1340 of 550 hp. As the AT-6 (USAAC and later USAAF), SNJ scout trainer North American (US Navy), Harvard (RAF and British Commonwealth air forces) and Wirraway light attack aircraft with twin forward-firing guns and one on a swivel mount in the rear cockpit (RAAF), it was built in large numbers in many countries and provided advanced flying training for more Allied airmen than any other model.

It initially appears to be coincidence, but is probably more a measure of the company, that three of North American's designs became standards by which other aircraft were judged. The AT-6 series was a resounding success and saw production of some 15 000, and the Mustang, made in similar numbers, was one of the best all-round fighters of World War II with its Rolls-Royce Merlin engine made under licence in the USA by Packard. The F-86 Sabre was also acknowledged as one of the finest jet fighters of its time, and saw active service during the Korean War. An example has been imported by Warbirds from the Philippine Air Force, but its condition on arrival was worse than anticipated and it is unlikely to fly again.

Wartime service pilots received their initial training in biplanes — Tiger Moths or Stearmans, according to nationality — but went on to learn about such things as retractable undercarriages, variable-pitch propellers, flaps and even closed cockpits in the large, robust, all-metal Harvards. Once the pressures of wartime training had eased off and things could be done at a normal pace, the RNZAF also used its Harvards for regular force primary training at Wigram, until they were replaced from mid-1976 by the Hamilton-manufactured Aerospace CT/4B Airtrainers.

Changes have been made among the Harvard pilots at Ardmore, too. The aircraft have always been in demand for flying displays at airshows around the country (and usually succeed in upstaging their modern replacements, flown by the RNZAF Red Checkers team); and a Warbirds routine has been highly popular from the late 1970s.

Initially the display consisted of a pair of Harvards doing aerobatics and low-level beatups, usually flown by Trevor Bland and Ernie Thompson in ZK-WAR and ZK-ENE. Development of the theme has seen a four-part formation team entertaining the crowds with aerobatics and

The noisy end of NZ1065 over South Auckland.

Opposite: NZ1066 was one of the first Warbird Harvards.

EN E
NZ1066

Still in its 1950s training colours, NZ1025 was the first Harvard to fly in civilian guise, after being bought from the RNZAF by South Canterbury farmers Dave Diamond and Jim Sullivan. Now owned by Warbirds' vice-president John Greenstreet, it is seen over the Hunua Ranges, not far from its Ardmore home base.

A buzz-and-break is the usual way the formation team rejoins the circuit.

a synchronised routine, the pilots being ex-RNZAF personnel with plenty of Harvard and display experience.

But the emphasis was changed for the 1988 season. Apart from the formation leader, Keith Skilling, who is both ex-RNZAF and NZ Warbirds chief instructor, the team members are all aero-club-trained pilots who have progressed from their Cessnas and small aeroplanes up to something more substantial. Rob Booth, John Greenstreet and Steve Taylor are all experienced aerobatic pilots, but their training into a tightly knit formation team represents something of a milestone for the Warbirds.

After all, if there's a continuing flow of new pilots wanting to fly such things as Harvards in a very public manner, it bodes well for the movement's future.

John Crosbie's blue Tiger Moth against the blue of Lake Wakatipu.

DH82A TIGER MOTH

As if open cockpits aren't breezy enough, Rusty Butterworth's airshow act is on the wing of Tony Renouf's Tiger Moth.

To the de Havilland Tiger Moth, the very symbol of the vintage aeroplane in New Zealand skies, goes the distinction of training more British Commonwealth air force pilots to fly than any other single type in the days of the Empire Flying Training Scheme during World War II. So although it may not have gone around as a warbird in the strict sense of firing guns in anger, the Tiger Moth deserves its special place in the ranks of the Warbirds.

Most of the wartime pilots remember it with some fondness, although many at the time cursed its idiosyncrasies — the lack of wheel brakes, for which the tailskid could compensate to only a certain degree, the poor visibility from the rear cockpit on the ground in the three-point attitude, the strange control harmonisation that demanded full and correct use of the rudder to stop it skidding all over the sky, and the amazing number of draughts of chill air that penetrated even the bulkiest winter flying clothing in the frosts of an Otago winter's morning.

However, once they had learned its tricks they grew to appreciate this little biplane, even as they progressed to the Harvard advanced trainer and from there to fighters or twin-engined trainers on their way to the bombers of Europe. Out of a total of some 8 500 produced in several countries, 345 Tiger Moths came out of the de Havilland factory at Rongotai, Wellington, from the end of 1939, to help train air force pilots for their part in the world-wide conflict.

Its origins date back to 1925, when the first of the wood-and-fabric Moth family of open-cockpit two-seaters was made by the de Havilland factory and soon became the standard British light aeroplane. By 1931 the Moth had grown a welded steel-tube fuselage structure and interest was shown by the RAF. As a trainer, however, it was handicapped by the positioning of the top wing directly over the front cockpit, with the struts seriously impeding access for the instructor who might just want to depart in a hurry and take to the parachute which was hampering his movements.

The answer was to move the top wing forward, giving all wings sweepback to preserve the centre of lift, and so was born the classic Tiger Moth with its own DH82A type number but sharing most components with others in the growing Moth family. Hundreds are still flying in various countries today, with perhaps 30 airworthy in New Zealand.

One of the most recent Tiger Moth rebuilds is ZK-ASG, seen *(opposite)* over the Hawke's Bay countryside near Waipukurau. Its 1950s RNZAF paint scheme contrasts with Russell Broadbent's Tiger Moth *(above)* which shows bright civilian colours against part of Auckland Harbour.

Their survival rate is testimony more to good original design and repair work than careful flying, for the Tiger Moth played a very big part in establishing the country's aerial topdressing industry and its agricultural survival. The work was hard and the aeroplanes unsuited to the task, but they were cheap and in ready supply, and if the excessively heavy loads they were obliged to carry sometimes led to them falling off the ends of runways, there were always plenty more parts back in the hangar to repair them and keep the topdressing fleet in the air.

After a few years of such misuse, the surviving Tiger Moths were replaced by aircraft better suited for the job, and they retired to a more gentle life of towing gliders, while the country's aero clubs still used some for training, especially in aerobatics. By the end of the 1960s most had been retired from even that sort of use, dismantled and stored in hangars and farm sheds, although a few staunch enthusiasts were still to be found enjoying their elderly biplanes. Sometimes they were scoffed at by those who had graduated, as they saw it, to the comfortable, modern cabin light aircraft, invariably American in manufacture and easy to fly, with lots of instruments and radios to make them efficient transport machines.

But the wheel has turned full circle, and those who have remained faithful to their trusty vintage biplanes have seen them become respectable once more. Appreciated for their individuality, a quality not found in the rows of look-alike Cessnas and Pipers which line today's aero club fields, the Tiger Moths have mostly needed full rebuilds to bring them back to 'as new' condition, and are owned and flown by an interesting mixture of pilots.

Among the Warbirds members, some are of the younger breed who learned to fly in modern aircraft but like something with a bit more character and challenge — not to mention aerobatic capability. Others have had their Tigers for years and have seen respectability grow up around them again, while one or two topdressing pilots, both past and present, are reliving the challenges of younger years. Another category learned to fly in Tiger Moths long ago in aero club days, and unashamedly admit to nostalgia, and there's even one tiny minority who learned to fly in the RNZAF during the war and can still turn a tidy barrel roll.

Most of the Tigers owned by Warbirds members are resident at Ardmore — where indeed there are more to be found than at any other single spot in New Zealand — but several are based in the South Island, as far away as Lake Wanaka. Some have been painted in military colours and markings recalling their original use (although it comprised the shortest phase of their history), but the majority are in mufti and make a colourful spectacle at any airshow.

Tiger Moth pilots after a formation flight to commemorate Air Force Day. From left: Russell Broadbent, Ross Duncan, John Crosbie, Tony Renouf, Bob Melse and Syd Broadbent.

Tiger Moths are often found in flocks.

STEARMAN

Lloyd Stearman's famous training biplane filled exactly the same role for the Americans as the Tiger Moth did for the British fledgling pilots of an entire military generation. Both were originally produced around 1930; and were employed as primary trainers during World War II (and, in several countries, for some time afterwards). Around 8 500 of each were produced by a variety of manufacturers, and both types moved on to agricultural flying when their military days were over, after which they retired into private ownership for sport flying.

But the Stearman, being American, is bigger than its British contemporary. Not much bigger, just a few centimetres in all directions, with a fuselage of generous cross-section so that the overall impression is one of beefiness, whereas the Tiger's is of slimness and delicacy.

The Americans were fond of their big round engines, too, and didn't mind having them uncovered out in front where everybody could see what fine pieces of machinery they were. And if the drag factor from an uncowled radial engine was a bit much for the 1930s biplane fighter, it didn't matter the slightest on a trainer which was seldom used for going in straight lines anyway. A big engine was needed to haul the Stearman's considerable maximum weight of more than 1200 kg around the sky (strength comparisons were inevitably made with the Brooklyn Bridge, and nobody ever seemed to bend one in aerobatic flight) and it carried on a fine tradition of no-nonsense biplane trainers in the American manner.

Only one Stearman has seen active service in New Zealand, imported only a few years ago by Len Cowper, a Kiwi who has spent considerably more years flying in Asia than anywhere else. He resigned from the RNZAF while in Singapore in 1958 to go flying with Cathay Pacific, then a bush airline operating one DC.3 and a DC.4 from the small airfield of Kai Tak, Hong Kong, and finally retired from an executive position with the airline in 1982. Since then he has established a business in Hawaii, selling New Zealand pre-cut homes, as well as a flying school and rental/charter operation, although his biplane is permanently resident at Ardmore and sees only occasional flying at suitable airshows.

The Stearman still flies under its United States registration of N4036, which helps avoid some of the problems with first-of-type certification in this country. Used as a basic trainer in the USAAC, it was one of 25 given to the Philippines in 1946 to help rebuild their air force after

Ann Barbarich tries the Stearman as a skydiving platform.

Len Cowper puts some muscle into starting his Stearman *(left)* and then pulls the handle to engage the inertia flywheel *(above)*.

the Japanese occupation, and declared surplus 10 years later. All were bought as a job lot by a Filipino mechanic for $US135 each, and various combinations of engine, fuselage and wings were used for pineapple crop spraying.

This particular bundle of hardware was bought by an American who had it fully rebuilt in Manila over a period of two and a half years. According to its present owner, it will later be re-covered and repainted in its original Army Air Corps training colours of blue fuselage and yellow wings.

The tendency in the United States today is to re-engine Stearmans with bigger and more powerful engines to improve performance and make them more spectacular in aerobatics. N4036, however, retains its original type of 225 hp Lycoming R-680-17 9-cylinder radial, and while its performance is stately in such things as rate of climb, Len Cowper's aerobatic routine takes full advantage of its good points and shows how little sky it really needs for manoeuvres.

Its engine shows one very big difference from the 130 hp Gipsy Major found in the contemporary Tiger Moth. Whereas the British machine is started by swinging the propeller in what is inevitably known as the Armstrong method — by hand — the Lycoming is fired up with a hand-cranked inertia starter.

That involves inserting the crank handle into a socket in the fuselage side just behind the engine (after setting brakes and throttle and switches in the cockpit, of course) and, from a precarious position on the wheel, winding laboriously to bring the flywheel up to speed. Once the winder is worn out or thinks things are going fast enough, the crank is removed and a toggle pulled to engage the flywheel and turn the engine over. All going according to plan, the engine starts in a cloud of reassuring blue smoke and ticks quietly away to itself. From time to time, however, things die away to an uneasy silence and the aspiring and perspiring aviator has to repeat the procedure; twice in succession is about all that can be managed at any one time before the pilot has to hand over to somebody else.

In recent years the Stearman has shown an increasing reluctance to start, and the firing-up procedure has been accompanied by a strange lack of knowledgeable onlookers, most Warbirds members suddenly finding important things to do on the other side of Ardmore. The owner, after many fruitless attempts before giving up and resorting to an electrically driven starting device, finally came to the conclusion that the magneto was not strong enough to spark at starting speeds, so recent work on that should see a minimum of worn-out helpers draped over nearby wings, recovering their breath.

The L4B is based on an airstrip near Coroglen on the Coromandel Peninsula.

PIPER CUB

Real warbirds with histories of live action in war theatres are not particularly common among Warbirds aircraft, despite the military roundels worn by most. Even the Harvards, big and noisy as they are, have seen service merely in the training role and have never been fired upon in anger. Only the fighters have ever fired their guns — and those only at practice targets.

But one outwardly innocuous little aeroplane has had a career rather more varied than most, and in exotic surroundings. Not too often seen these days away from its base on a little airstrip near Coroglen on the Coromandel Peninsula is a Piper L4B Grasshopper, the military version of the ubiquitous Cub, used for artillery spotting and general liaison work involving light aircraft wherever the United States military forces were active during World War II.

The J-3 version of the Piper Cub was produced in larger numbers than any previous light aircraft. Some 20 000 came off the assembly lines in the 10 years after 1937, so it's not surprising that most general aviation airfields in the United States of America have at least one Cub, usually painted in the standard yellow colour scheme. Included among that production were 5 673 wartime L4B Grasshoppers, naturally enough painted in military olive drab and bearing USAAF stars and bars, and with an immediate recognition feature in their extra length of cockpit transparency for visibility behind and below.

The Warbird example was built in August 1942 for the United States military forces, and after seeing its share of wartime service was bought by the British Solomon Islands Government for aerial spraying in malaria control work. In 1947 it was bought by two RNZAF men and taken to Fiji for a full overhaul, and two years later found its way to New Zealand and was registered ZK-ATU. During an active life, which included modification in 1953 for aerial photography in the Wairarapa, its identity reverted to that of an ordinary J-3. Eventually its condition had deteriorated so much that it was no longer flown, and it went into long-term storage in the mid-1960s.

However, in 1983 it was taken to Jim Renner's airstrip at Coroglen for a full strip and rebuild by six enthusiasts, emerging one year later in its former guise as a genuine warbird. Somewhere along the line its original 65 hp Continental A65 was replaced by the more spritely C90 which, combined with a reduction in empty weight to just 670 lbs, gives it sparkling performance.

Also re-engined is a second J-3 — but genuinely civilian — Cub, based at Ardmore along with most of the other Warbirds fleet. ZK-AHC is one of three J-3s imported in 1939, and all are still airworthy in various parts of New Zealand, having escaped the fate of RNZAF impressment that overtook most private aircraft at the beginning of World War II. Originally powered by just 40 ponies, ZK-AHC has been through a succession of Continental transplants and now sports no less than 85 hp.

Flown for 30 years by a succession of aero clubs and private owners, the J-3 was tipped on its back by a gust of wind at Palmerston North in 1970, but two years later was bought by the Davis-Goff brothers of Masterton and given a full rebuild. It emerged in 1976 in even better condition than new and spent some years in the Wairarapa before being bought by Jim Peace and moving north to Ardmore.

A third Cub variant is in Warbird ownership, but is not expected to fly before the end of 1989. Abe Lincoln has imported an ex-US Navy J-5 ambulance model from Hawaii and is in the process of giving it a full rebuild. When finished it will sport the registration ZK-USN.

Not many New Zealand warbirds have served overseas in wartime, but ZK-ATU saw active service in the Solomon Islands.

Opposite: ZK-AHC is one of three J-3 Cubs imported new in 1939 and still in good flying trim.

Austers come in a bewildering variety of model numbers and names, but only the expert can tell the difference between the J/IN Aiglet in the foreground and the J/5 Adventurer. Both use the ubiquitous DH Gipsy Major engine also found in the Tiger Moth.

AUSTER

Although the thoroughly British Auster still seen in New Zealand in some numbers is a peacetime product of the 1940s and 1950s, its origins were very much influenced by World War II. It all started back in 1928 with two English brothers, C. Gilbert and Gordon A. Taylor, making a light tandem two-seat runabout aeroplane in the USA and intending to set the aviation market on its ear. A couple of company failures later, the product had evolved into the E-2 Cub under the auspices of William T. Piper, a wealthy Pennsylvania oilman, who continued to make it from 1935 as the J-2 and later J-3 Cub, the aeroplane which taught a whole generation of Americans to fly.

Meanwhile C.G., the surviving Taylor brother, had gone ahead on his own and developed the Taylorcraft Model A, a side-by-side two-seater which was imported into England in 1938 and set up for local production under licence. By the time 20 were sold during the summer of 1939, people's minds were on other events in Europe and the project was quietly shelved. However, the English Taylorcraft company kept going with contracts to repair damaged fighter aircraft, something in plentiful supply in the years to come.

But the little two-seater was not forgotten and, re-engined with a Blackburn Cirrus Minor I of 90 hp, was chosen as an artillery spotter to see action in France up to the Dunkirk evacuation. Before major production could begin in 1941, however, it had to be given a generic name in line with all RAF aircraft. So in keeping with the contemporary trend towards Hurricane, Tempest, Tornado and other strong winds, it became the Auster, the Roman word for a warm southerly wind and more in keeping with its gentle performance. Several hundred had been produced by the end of hostilities, many of them seeing civilian service in the following years — including several in New Zealand — and Auster Aircraft Ltd naturally adapted its well-proven design to peacetime production for the following 15 years.

Further military models were developed, becoming rather ungainly and heavy and powered by a succession of bigger engines, until in 1961 the AOP (air observation post) role was taken over by the helicopter and the Auster company, its lucrative military contracts gone, began a rapid slide into oblivion.

In New Zealand, the RNZAF used Austers for forestry patrol and army liaison work. Instead

An Auster's forte is its ability to fly slowly, helped by the generous wing area.

of the military model, however, the RNZAF chose the civil J/5, of which six were ordered in mid-1947 as airframes only and fitted with Gipsy Major engines surplus from the wartime Tiger Moth production at Rongotai. They served with a number of units in various places around New Zealand, gradually reducing in numbers through being bent or sold, until the last retired to civilian life in 1970. A seventh Auster, bought in 1956, was actually a genuine military T. 7C, and served with the RNZAF Antarctic Flight on the ice until 1959. It is now at Wigram with the RNZAF Museum.

Two Austers are owned by Warbirds members, and although neither of them actually saw military service with the RAF during the Second World War or the RNZAF in later years, the type is recognised for its contribution to the military activities over a long period. A third Auster, regularly flown by several Warbirds members but actually owned by the Vintage Aero Club and based at different aerodromes around the country in rotation, is J/1 ZK-AUX, which holds the distinction of being the first civil aircraft to land on an aircraft carrier. As G-AERO and used by *The Aeroplane* magazine as company runabout, it landed on board the new aircraft carrier HMS *Illustrious* in October 1946 and repeated the exercise in May 1948 aboard the Canadian light fleet carrier HMCS *Magnificent*.

A clue to the Auster's observation origins is its large areas of clear vision panels.

From some angles the Issacs Fury needs only a 7/10 scale pilot to look entirely accurate.

REPLICAS

A category of aeroplane quietly growing in numbers in the Warbirds movement has characteristics all its own, and cannot be compared directly with military fighters and trainers. These aircraft did not emerge from established factories and have never been used by air forces at all; yet their shapes are familiar to those who read aircraft history books and they occupy a small but significant part of the overall warbird scene in New Zealand.

These are the replicas, built in recent years by enthusiasts in home basements and garages to plans adapted from the original types, which certainly saw service with air forces in Europe long ago. Three types — Isaacs Fury, SE5a and Fokker Triplane — are owned by Warbirds members and take an active and popular part in airshows.

The first to appear in New Zealand skies was the Isaacs Fury, developed by Englishman John Isaacs as a $\frac{7}{10}$ scale version of the elegant Hawker Fury biplane fighter of the 1930s. The single-seat Fury gave rise to a number of similar-looking aircraft powered by versions of the same Rolls-Royce Kestrel V-12, ranging from light bombers to navalised Nimrod and Osprey fighters.

The aircraft was renowned at the time for its pleasant handling qualities, which the designer went to some lengths to duplicate in his homebuilt version — not an easy task, for merely scaling an aeroplane down doesn't necessarily mean that all its original characteristics will be retained. For ease of construction he gave it an all-wooden structure, covered in fabric, and the first prototype flew in 1963, powered by an inverted four-cylinder Walter Mikron engine.

The Fury II, which became the accepted design built in many countries, can use any power plant from 65 to 125 hp, and the majority are equipped with the readily obtainable American horizontally opposed four-cylinder engines of around 100 hp. Even if they don't sound nearly as impressive as the original Rolls-Royce versions, they do have good performance in take-off and climb and are good aerobatic aeroplanes.

Three of the four Isaacs Furies so far built in New Zealand are owned by Warbirds members and based around Auckland. Colin Glasgow's ZK-DMN shares an Ardmore hangar with a wide assortment of aeroplanes which include a Fokker Triplane and an Auster, while David Simpson's Fury ZK-RFC is based at Dairy Flat and has been equipped with brakes in deference to the

The Fokker Dr1 Triplane has more wings than any other aeroplane in New Zealand.

The original SE5a was one of Britain's most successful World War I fighters.

Far from the war-torn fields of Europe is Keith Trillo in a full-scale Fokker Triplane replica.

narrow runway there. A group of six owns ZK-JHR and operates it from Hobsonville, a most appropriate aerodrome with its wide expanse of grass and its continuous military use since 1928, when the original Hawker Fury was in production.

Rather earlier in initial design is the SE5a, one of the better-known (and respected) fighters of World War I. A handful of these aircraft, powered by 200 hp Hispano-Suiza or Wolseley Viper engines, are still in existence around the world, but the Canadian company Replica Plans first flew its modern-day $\frac{7}{10}$ scale SE5a in 1970. Like the Fury, it uses easily contructed wood-and-fabric techniques and modern American engines.

Two SE5a replicas are flying in New Zealand. The warbird, based at Ardmore by its group of owners, was built by Stuart Tantrum with a Lycoming O-235 engine and fitted with dummy guns and laced-up panels on the fuselage, with fine attention to detail.

Also built by Stuart Tantrum, in conjunction with John Lanham, is an even more ambitious replica fighter project. The Fokker Dr1, usually known as the Triplane, is no scaled-down miniature but is a full-sized exact replica of the original made famous by Baron Manfred von Richthofen, the near-legendary Red Baron.

Some concessions have been made in such areas as modern steel tube sizes in place of the original metric measurements, and a radial engine has been fitted in place of the proper (and very difficult to obtain) 110 hp Le Rhone rotary. Otherwise the Fokker is completely authentic, even down to the 76 ribs used in making the three sets of wings, and it is one of the most distinctive shapes in the skies.

One notable area of authenticity is the plane's flying characteristics. Fighter aircraft of 70 years ago were made unstable for easy and quick manoeuvrability, and ZK-FOK displays all the original quirkiness, so unlike anything else its present owners have ever flown. They describe it as 'Well. . .different', as the slightest relaxation on the part of the pilot will result in its flying sideways or dropping off into a spiral dive.

But that's one of the attractions of a replica fighter. How else is a modern pilot going to appreciate what his ancestral aviators went through?

Opposite: a diverse flock of fighter replicas approaches Ardmore home base.

It could be a wartime coastal patrol, beside Kapiti Island's rugged western face.

DH89B DOMINIE

T he British aircraft industry during the 1930s was slow to adjust to the principle of stressed-skin, all-metal construction that was being appreciated by manufacturers in Germany and America. The latter country in particular was showing the way in modern airliners, and in 1933 Douglas produced the DC.1, made completely of aluminium and with not only variable-pitch propellers but also retracting undercarriage. This led to the DC.3 a couple of years later. Other airliner companies followed suit, notably Lockheed, whose smaller aircraft were seen on the New Zealand main trunk routes later in the decade.

Typical of the British manufacturers was de Havilland, which in the same period was still steadfastly producing medium-sized aircraft made entirely of wood covered with fabric, and powered by one or more variants of its own trusty Gipsy engine with four or six cylinders. The advantages of wood were usually stated as including its light weight for a strong structure, and certainly for the modest speeds of de Havilland's aircraft there were no aerodynamic problems. Adherence to the biplane principle, for its larger passenger-carriers at least, gave high drag with all those struts and wires, and the nearest the company came to retracting the undercarriage (apart from the specialised Comet racer that took line honours in the London–Melbourne Air Race of 1934) was enclosing the wheels, grudgingly no doubt, in fairings swept down from the engine nacelles.

But de Havilland made aeroplanes that looked right, even if they could not match their trans-Atlantic counterparts for speed. Another DH product in that gruelling race half-way round the world was the newly developed DH89 Dragon Six, which combined features of two slightly earlier twin-engined biplanes into one elegantly shaped machine for carrying up to eight passengers. One example, ZK-ACO, was flown by the all-New Zealand crew of Jim Hewett, Cyril Kay and Frank Stewart, to finish ninth overall and fifth in the handicap section after five extra days spent repairing damage caused by taxying into a Cloncurry fence in the dark.

The Dragon Six became the Rapide, a popular feeder airliner of the 1930s, five of which reliably served in New Zealand on the Cook Strait run between Wellington, Blenheim and Nelson. When war broke out they were impressed into RNZAF service, first as communications aircraft and flying classrooms for trainee observers, and later on they were used for general reconnaissance duties in Fiji, complete with bomb racks. Hurricanes and general tropical conditions led to the end of that group of planes, but wartime production of the improved DH89B Dominie saw more of the type serving in both Fiji and New Zealand, many of them with No. 42 Squadron on scheduled services around the North Island bases.

Six were transferred to the fledgling NZ National Airways Corporation when it started operations in 1947, while a seventh, the sole surviving pre-war Rapide, was acquired through the merger of the West Coast airline Air Travel into the main corporation. One was lost by fire at Rotorua in 1950, and the others were progressively sold to third-level airlines from the mid-1950s, the last (ZK-AKY) going to Ritchie Air Services for Fiordland tourist flights in 1964.

One Dominie is in the Warbirds fold. ZK-AKU went to Nelson Aero Club in 1964, and later to Rotorua Aero Club, where it operated two scheduled flights a day for two years. It was bought by David Gray in 1968 after a short period on the South Island tourist route from Christchurch. A high-time aeroplane, it has flown more than 20 000 hours and today still sees plenty of activity as a privately owned airliner, its seating capacity called upon to transport enthusiasts to distant airshows where it performs.

But life since retirement from airline duties has not all been ease, for ZK-AKU took part in the 1969 London–Sydney Air Race in parallel with that other Kiwi effort 35 years previously. What made the later performance all the more praiseworthy, however, was the fact that the three-man crew of David Gray, Jack Moon and Ray Cooney had to fly the Dominie all the way to London before they could even begin the race. Problems *en route* delayed them and they arrived at the Gatwick start on the day of the race, only hours before having to turn around and fly home again.

The vintage section was cancelled because the number of entries was too small, but ZK-AKU finished in Sydney in 22nd place overall, the only vintage type to qualify and also the oldest aircraft to have flown the Tasman Sea. The DH89 total elapsed time for the England–Australia race beat the previous record by 110 hours, and the Dominie remains the only vintage type to have flown from New Zealand to the UK and back.

Tawaka (grey duck, its NAC name) is now resident at Dairy Flat on Auckland's northern outskirts. After more than 40 years of civilian identity it has reverted to wearing a military colour scheme. David Gray and Harry Norton have researched the original wartime paint of their Dominie, which now complements the other example of the type still flying in New Zealand, ZK-AKY, which is based at Masterton with the NZ Sport & Vintage Aviation Society and painted in its original NZNAC colours.

Opposite: this Dominie has been in private ownership for more than 20 years.

NZ 528
ZK-AKU

The Devon is unique in carrying both an RNZAF-style colour scheme and Warbirds identification.

DH104 DEVON

The de Havilland company's first post-war civil design was intended to replace the 10-year-old Rapide, or Dominie, as a feeder airliner carrying eight passengers. The new Dove, which first flew in September 1945, shared very few characteristics with its predecessor, however. Certainly it had two engines named Gipsy Queen, just like its pre-war stablemate, but they bore little resemblance to the earlier Queens, being supercharged and each producing almost twice the power at up to 380 hp.

And although its tail shape was still recognisably de Havilland, the new Dove was thoroughly up-to-date and more than 10 years in advance of the Dominie. It was the first British transport aircraft to have fully feathering and reversible pitch propellers, but more immediately obvious was its monoplane configuration, two elegant wings spanning more than 17 metres and made entirely of metal, as was the fuselage. On the ground it sat on an undercarriage which was not only retractable but also boasted a nosewheel in the most modern fashion, and to round off the impression of a workmanlike small airliner was a cockpit for two pilots forward of the passenger cabin and on a higher level, with a door to shut the aircrew away from hoi polloi down the back.

The military version of the Dove was named the Devon, and became familiar in New Zealand skies as the RNZAF bought 30, taking delivery of the first in 1948 and phasing out the remaining survivors 30 years later. Most of them were bought in 1952 as replacements for Airspeed Oxfords (as twin-engine trainers), and Avro Ansons (as navigational and signals trainers), and Avro Consuls and de Havilland Dominies (as communications aircraft). They proved good workhorses and were popular with the citizens of southern Christchurch near RNZAF Base Wigram, being whisper-quiet by comparison with the Harvards.

In the course of their military service, seven met with accidents which caused irreparable damage. Two were given to the Royal Malaysian Air Force in 1968, and four were sold to civilian buyers in Australia not long afterwards. The remainder were retired as their navigational training role was taken over by surplus NZNAC Fokker Friendships, and Cessna Golden Eagles replaced those being used for communications. Several were sold to local operators over a period of some years.

NEW ZEALAND WARBIRDS
08

Two examples went in a bulk tender in November 1984, complete with a large quantity of spares. They spent five flightless years outdoors at Woodbourne, Blenheim, but had been properly inhibited and needed a minimum of work to be readied for temporary civil certificates of airworthiness, following which they were ferried to their new homes for more permanent conversion and refitting for passenger comfort. ZK-RNG went to new owners in Palmerston North for a long period of refurbishing, while ZK-KTT was bought by a syndicate and is often flown between its Whangarei base and Auckland.

Towards the end of 1988 the northern Dove once more became a Devon, being bought by a group of 30 Warbirds members. ZK-KTT is now based at Ardmore, back to its old number of NZ1808 and painted in a colour scheme similar to the VIP Devons operated by the RNZAF but sporting a New Zealand Warbirds identification above the cabin windows.

Carrying a total of 10 people, it is used for carrying groups of Warbirds members to distant airshows and other appropriate occasions, something not possible with the single-seat fighters or even two-seat Harvards. After all, it is the members who make up any such organisation, and moving them around the country is an important part of the overall scheme.

The DC.3's RAF wartime paint scheme combines both a military look and reasonable visibility for Ardmore's busy circuit.

DC.3

One of the most enduring pieces of aviation history is the Douglas DC.3, that friendly old lady of the skies which first flew more than 50 years ago.

Developed as a transport aircraft in 1935 for carrying 14 sleeping or 21 daytime and upright passengers in hitherto undreamed-of comfort and speed, it later happened to be exactly the right aircraft for wartime military cargo and troop-carrying transport. More than 10 000 were made in a 10 year period, and it looks as though the DC.3 will be flying for still more generations.

The original Douglas Commercial, or DC.1, was a trend-setter with structural and aerodynamic features found on only one or two of its contemporaries — and on very few military aircraft of 1933. It boasted an all-metal aluminium stressed-skin structure, variable-pitch propellers on its radial engines, and retracting undercarriage, all features which were carried on the production model, the 14-passenger DC.2. An excellent demonstration of its comfort and speed was the fact that it took second place in the 1934 London–Melbourne Air Race while carrying fare-paying passengers and mail, finishing only hours behind the de Havilland Comet, a racing model.

Given more powerful engines, along with a longer fuselage and wings, the DC.2 became the immortal DC.3, one of the great success stories of aviation and entered airline service in June 1936. The wing roots, however, remained the same, leading to that well-known wartime story of the DC.2 with a bomb-damaged wing which was fitted with the corresponding unit from an equally unairworthy DC.3 — and flew.

Some 800 DC.3s were being flown by the airlines a few years later when war broke out, and the aircraft entered military production as the C-47, a general cargo and troop carrier. A unique combination of robust construction, load-carrying capability (many stories tell of inadvertently overloaded aircraft still delivering their cargoes) and ease of flying and control made it popular with aircrew of many Allied air forces, not least among them the New Zealanders around the Pacific.

The RNZAF took delivery of the first of its 49 C-47 Dakotas in March 1943, and soon put them to use in support of fighting units in the Pacific Islands. They were still busy at the war's end in 1945, repatriating servicemen and prisoners of war, and also performed the role of couriers to occupied Japan and the UK with No. 41 Squadron, although the long-haul flights were taken over from the early 1950s by Bristol Freighters and Handley-Page Hastings transports.

The other squadrons flying C-47s were No. 42, which retained the last of them as VIP transports until they were finally retired in 1977, and No. 40, which operated a number in a quasi-civil role as a precursor to the NZ National Airways Corporation. Most of the aircraft and aircrew went from the squadron into the airline when it officially started in April 1947, and NZNAC flew a total of 29 DC.3s, all but three of them bought from the RNZAF, until the last was sold in 1973.

But that wasn't the last of the type's operations in New Zealand. Mt Cook Airlines used three on its tourist routes south from Christchurch for many years until they were replaced by modern Hawker-Siddeley HS.748s, while South Pacific Airlines of New Zealand (SPANZ) tottered along for just over five years from 1960, using three Australian-purchased DC.3s in direct competition with NZNAC before its demise.

The DC.3 has also played its part in the local aerial topdressing industry, with 13 having dropped superphosphate in five-tonne lots on the countryside. While the smaller aircraft have used farmers' local airstrips, the DC.3's operations have tended to be in areas lacking such amenities but not too far from a major airport with a runway long enough to take a fully laden ex-airliner.

The last of those, based at Gisborne, has now been retired, but the sonorous drone of twin Pratt & Whitney radials is still heard in local skies, since two DC.3s are operated as freighters by Classic Air Services.

Making really sure that the model stays around for a while yet is a group formed under the Warbirds' banner, which in 1987 imported a 1943 DC.3 from Queensland. It has seen most of its 43 800 hours in the Far East, with the USAAF in that theatre for its first 10 years, after which it was transferred to Philippine Airlines, later finding itself on airline service in Papua New Guinea in 1970, and with Bush Pilots Airways (later Air Queensland) two years later, before going to Mackay Air Museum.

Not a particularly high-time DC.3 (one with 102 000 hours under its wings is still in use in New York State on commuter services), the Warbirds' ZK-DAK was the subject of much debate over its colour scheme. Should it be olive drab camouflage, like the RNZAF's many C-47s which the group aims to commemorate? What about the visibility aspect in the busy circuit at Ardmore, with surrounding hills just that colour? Should it be painted in the RNZAF VIP colours, much

50

prettier but still preserved on one of the air force's original such aircraft at Wigram? In the end it was decided to paint the DC.3 in a compromise scheme of olive drab with D-Day invasion stripes for visibility.

A 12 712 kg airliner is not the sort of thing the average aero club type is let loose in, even if he or she does have a tailwheel rating. Its weight — not to mention those two big Pratt & Whitney radial engines — takes it right out of private pilot territory, and those qualified to fly ZK-DAK tend to be current airline pilots, some with several thousand DC.3 hours.

But there is a new trend discernible with this particular warbird, one of the milestone aircraft in aviation history. Some pilots converting to the type are of a new generation which has not been brought up on DC.3s — and a welcome number of those are women who have graduated into airline flying and want to extend their flying experience.

ZK-MUH

DHC-1 CHIPMUNK

The origins of de Havilland Canada's endearing little trainer go back to 1943, when production of the Mosquito for the European war effort was under the control of the Canadian government and thoughts were turning to post-war projects. A rugged bush plane was suggested, larger than the pre-war Fox Moth design, that would use many Tiger Moth parts; and the company felt that demand would warrant the undertaking of every aspect of design and production.

In England, however, the immediate need for a trainer to replace the venerable Tiger Moth was considered to be more urgent. As well as that, a two-seater project would make an ideal engineering exercise at Downsview, Ontario.

Work on the DHC-1 Chipmunk began in 1945; the bush aircraft came later in the form of the enduring Beaver. Six months after its first test flight in May 1946, the Chipmunk was shipped to England. Adopted by the RAF as the Tiger Moth replacement, it was mostly made in England, with only 218 coming off the production line in its home country of Canada.

Although the Chipmunk bore little outward resemblance to the Tiger Moth — all-metal construction and monoplane configuration were the most obvious differences — the pilot felt immediately at home. In front of him were familiar instruments and a joystick, and in flight the sound of the Gipsy Major was just the same, even if he was shielded from icy winter blasts by a sliding canopy.

But the Chipmunk showed few of the Tiger Moth's quirks in flight, and quickly made a name for itself as a delightful aeroplane to fly. Control harmony and smoothness of response made it ideal for aerobatics and all over the world people have developed its airframe potential by installing bigger engines in the search for even better aerobatic performance.

In Australia more that 90 Chipmunks have been flown by the civil operators, but for some reason very few have migrated to New Zealand. Perhaps it was the high initial cost, particularly when compared with the price of surplus ex-RNZAF Tiger Moth which could do almost as much in flying training, but whatever the reason the aero clubs nearly all retained their earlier, fabric-covered, open-cockpit biplanes. And unlike the RAF, which adopted the Chipmunk as its standard trainer, the RNZAF retained the Tiger Moths for *ab initio* training until it was supplanted by the Harvard.

Opposite: the smaller warbirds can have recognisable pilots. John Wall rides in the back of his Chipmunk while Keith Trillo flies it from the front.

Three Chipmunks fly within Warbirds membership, all built in England and privately imported in later years.

ZK-DUC represents the Australian group, having flown with the RAF but later finding its way into the Royal Victorian Aero Club. There it was located by Ken Scott in the early 1970s and brought to New Zealand, later repainted in a colour scheme reminiscent of its RAF days. It has for some time been a familiar sight at southern airshows in the hands of present owner Simon Spencer-Bower.

The other two were imported direct from England and are based at Ardmore. ZK-TNR is an early example, passing through a number of RAF flying training units before being sold into private ownership. It briefly flew in Switzerland before returning to England and being shipped to New Zealand in early 1985, but flew only a few hours from Nelson until bought at the end of 1988 by a Warbirds group.

The most recent Chipmunk import is ZK-MUH, bought by Doug Dallison in 1985 and based by him at Cranwell until he returned to New Zealand with it three years later. Originally intended for the RAF, it served at Hamble, Southampton, with Air Service Training for BOAC/BEA flying recruits and, after a period of private ownership, was sold to the RAF College at Cranwell.

Always kept in hangars and treated well, ZK-MUH has for the past 20 years been flown only in good weather by experienced pilots. Present owner John Wall intends to keep up that tradition.

Ross Duncan in one of Warbirds' three Chipmunks, all of which were made in England although the design originated in Canada.

A big radial engine and long wing set the Beaver apart from lesser aeroplanes.

DHC-2 BEAVER

Most of the aeroplanes wearing military colours in the present Warbirds fleet have actually done so while in air force service, but there's an exception to every rule. In this case it is the bright red DHC-2 Beaver, a familiar sight around Ardmore and at airshows throughout the country.

Certainly the RNZAF operated a Beaver for a couple of years, but the example currently wearing roundels and the number NZ6001 has never actually seen military service. Instead, as with all the other civil examples in New Zealand, its background is aerial topdressing, but retirement some years ago into private ownership has given it the gentle leisure that it deserves after a most active life carrying freight in Canada, Africa and England, as well as dropping countless tonnes of superphosphate in this country.

The DHC-2 Beaver was the second design project of the Canadian de Havilland company at Downsview, Ontario (the first was that Tiger Moth trainer replacement, the Chipmunk). Intended right from the start to be the proverbial 'half-ton truck' for bush operations in Canada's vast inhospitable territory, it was a true utility aeroplane with a minimum of frills and fiddly bits that might break in service. A change of engine part-way through design stage resulted in the adoption of the 450 hp Pratt & Whitney R-985 Wasp Junior that gives the Beaver its sturdy, practical appearance, and the prototype first flew in August 1947.

More than 1600 were built during a 20-year production life. Thirty of those were imported into New Zealand for aerial topdressing until the late 1960s and, with such features as harmony of controls and the ability to lift a big load with safety, they proved very popular with their pilots. The first of those Beavers arrived in 1951 for Rural Aviation, the company having recognised the shortcomings of the topdressing Tiger Moths, which, although cheap enough to buy and operate, were not designed for agricultural work and met with a distressingly high number of accidents, most of them through overloading. Other operators quickly followed suit, with Fieldair heading the list at 19 Beavers flown over a period of more than 30 years.

Ironically, still more would have been used here but for a shortage of supply caused by a sudden rush of orders from another source. The USAF and US Army bought large numbers, which tied up the Downsview production line for some time, just as the Beaver was gaining acceptance in New Zealand service. It was a period of rapid expansion of the topdressing industry, so other types had to be found, and by the time the Beaver became readily available again, Cessnas and particularly the new Fletchers had filled the gap.

The RNZAF had its own Beaver for a short period, bought in mid-1956 by public subscription and used to support the Commonwealth Trans-Antarctic Expedition from December that year. Called *City of Auckland* because most of the funds had been raised there, it was given the serial number NZ6001 and flown as that until August 1959, when somebody pointed out that the number was not new. NZ6001 had previously been allocated to the Gloster Meteor, the RNZAF's (and New Zealand's) first jet aircraft, flown until 1950. The Antarctic Beaver then became NZ6010, but it was written-off after a whiteout accident on the Beardmore Glacier in January 1960.

However, because the RNZAF Beaver was known as NZ6001 for nearly all its operational life, the Warbird example has been given that number. Close inspection reveals the civil registration ZK-CKH, which it has worn since being imported in 1965 for Air Services Wairarapa and the customary aerial topdressing career, but the Beaver is now owned by an Ardmore-based syndicate and used purely for private and display flying. Apart from the extra registration letters, the paint scheme is an exact replica of that worn by the original Antarctic Beaver, the particular shade of red matching that on an original inspection panel.

It should not be confused with another RNZAF Beaver painted in authentic Antarctic colours, however. The RNZAF has recently bought another ex-topdresser for display, giving it the later, more correct serial number of NZ6010. But the Warbirds members thought of it first, and their Beaver is the more active. A newly overhauled Pratt & Whitney R-985 fitted at the end of 1987 should see it airworthy for many years to come.

This Beaver should more properly be seen against a background of ice and snow, but the Hauraki Gulf is now its normal habitat.

65
NZ1065
01
DGY

CT/4B AIRTRAINER

At first glance the Airtrainer may seem an unlikely candidate for warbird status. It is modern, with a tricycle undercarriage and none of the characteristics of the other aeroplanes in the fleet, such as big radial engines or tailwheels or high speed. It's also easier to fly than anything else — apart from perhaps the Cub or Auster — which might explain its popularity with Warbirds pilots.

But the CT/4B Airtrainer ZK-DGY has a very special place in New Zealand aviation history, for this aeroplane is the prototype of a locally designed and built military trainer that was sold to several air forces and is still in service as the RNZAF's *ab initio* trainer. Modern it may be, but the Airtrainer is just as much a warbird as any Harvard or Tiger Moth, the three types being responsible for training RNZAF pilots over the past 50 years.

Seating two pilots side by side, the CT/4B has a third seat behind for a passenger or observer. Although outwardly similar to the Airtourer and with the same wingspan, the Airtrainer is longer and has a 210 hp Continental engine, as well as a redesigned and restressed structure. Its systems are relatively complicated, as befits its use training air force pilots who will never fly simple aircraft during their entire careers, and its aerobatic capabilites have been shown off by the RNZAF's Red Checkers formation team over many airshow seasons throughout the country.

The origins of the CT/4B go back to 1960, with the Australian Victa company's newly formed aviation division working on a four-seat design by Luigi Pellarini, better known for his odd-looking Airtruk topdresser. Henry Millicer had won a Royal Aero Club design competition for a light aircraft with his two-seat trainer, which Victa took over and put into production as the Airtourer, the Pellarini project having been shelved. Lack of government support led to Victa's production stopping in 1966 after 169 aircraft had been made, but the following year Airtourers were being turned out in Hamilton on this side of the Tasman Sea.

Aero Engine Services Ltd had been formed in 1958 to repair and overhaul aircraft piston engines and, having expanded after some years of successful operation, took over Airtourer manufacturing. When the RAAF issued a specification for a new military trainer in 1970, AESL noted that it conformed closely in size with the Airtourer — and even more closely with the

Opposite: the RNZAF had standardised its red and grey colour scheme on its Harvards by the time the Airtrainer came along to replace them.

four-seat Aircruiser, the Millicer-designed prototype which had been acquired along with Airtourer production rights. The Aircruiser theme was quickly developed by AESL into a military trainer, and the prototype CT/4B Airtrainer, ZK-DGY, first flew in April 1972.

Developing an Australian design and selling it back to the Australians, against stiff competition from European companies, was something of a coup for the New Zealand team, and the RAAF order for 37 Airtrainers was soon followed by a contract for 25 from the Royal Thai Air Force.

The small production facilities at Hamilton Airport were suddenly inadequate, and the company was expanded accordingly.

AESL merged with Air Parts, also at Hamilton and producing the Fletcher topdressing aircraft, to form New Zealand Aerospace Industries. Further capital came from the government, represented by a 50 per cent shareholding by the two airlines, Air New Zealand and NZ National Airways Corporation, which also provided technical support.

A major sales drive was undertaken, in the course of which ZK-DGY became the most-travelled warbird, being ferried to the 1975 Paris Air Show, the prestigious showcase of the world's aviation industry. It appeared there in company with a Fletcher FU24-950, which also flew to Paris under its own power, and a BAC-built Strikemaster destined for the RNZAF, but of the trio of New Zealand aircraft only the Airtrainer was demonstrated at the show, by Dick Steele.

The RNZAF had already ordered a batch of CT/4Bs, and a contract seemed likely for the Danish government to buy 34 aircraft. Aerospace Industries had estimated the break-even point to be 100 Airtrainers to cover the expense of development and setting up the production line, and with some 80 having already been ordered and made for the Australian, Thai and New Zealand air forces, the Danish order would assure the company's and the type's future.

Alas, political pressure spelled an end to that, with the Swedes insisting that the Danes buy a rival Saab trainer. A further blow came in May 1976 with a refusal to grant an export licence for 14 Airtrainers, bought by the Swiss Breco Trading Company but rumoured to be destined for Rhodesia (now Zimbabwe), against which New Zealand had joined in a trade embargo. The embarrassment of that episode, and the possibility of similar things happening in the future, led to the company's management ending the CT/4 programme in August 1978 after a production run of 96 aircraft.

The prototype was bought by a Wellington pilot, Miles Nathan, who operated it for a number of years before selling it to a group of 40 Warbirds Association members.

Simpler and more straightforward, not to mention cheaper, to operate than a Harvard or

No other aeroplane is currently in both RNZAF service and Warbirds private ownership, although the ZK-DGY gives it away.

anything bigger, its aerobatic capabilities are finding favour with its owners and it is currently the busiest warbird based at Ardmore. The colour scheme closely matches that of the current RNZAF CT/4B fleet, which is understood to be regarded with disfavour by the military powers that be, who can imagine problems if civilian-flown misdemeanours are blamed on their own pilots.

Modern aeroplane though it is, with a design not yet 20 years old, the Airtrainer is fully deserving of its place in the Warbirds hangar at Ardmore.

Renowned as one of the best of all World War II fighters, the Mustang was not good at high altitudes until fitted with the superb Rolls-Royce Merlin engine.

MUSTANG

When a Mustang was seen and heard again in local skies in early 1985, it brought crowds to airshows and made headlines all around the country. Its pilot, Warbirds president Trevor Bland, instantly became the most envied pilot in New Zealand as he scorched low along an airshow runway at 300-plus knots or pulled up into a never-ending series of aileron rolls that took him out of sight into the summer sky, engine crackling with that distinctive Merlin sound. Every pilot or budding pilot on any airfield watching the Mustang would have given just about anything to be able to go flying in that superb fighter.

Just what is it about the Mustang that appeals so much? Is it the crisp shape with the squared-off wingtips and characteristic ventral radiator, or is it the incomparable sound of that V-12 Merlin, Rolls-Royce's contribution to the freedom of Europe 50 years ago? Without the Merlin engine the Mustang would not have made such a mark on the aeronautical world.

When the British Purchasing Commission, desperately seeking another front-line fighter in 1939, visited the United States of America to arrange the production of the P-40 for the RAF, it was sent along to North American Aviation. That company had been privately working on a high-performance fighter project, based on the Curtiss P-40 which had originally been designed with a ventral radiator. Combined with a laminar-flow wing and a number of its own existing systems it produced the Mustang.

Its main weaknesses of climb rate and poor altitude performance, caused by the original Allison V-12 engine installation, were cured instantly with the substitution of the Rolls-Royce Merlin. The Mustang went on to become arguably the best all-round Allied fighter of World War II, with its happy combination of performance, range, manoeuvrability and handling qualities. Even the Americans, their attitude towards it for some reason originally lukewarm at best, grew to appreciate the aircraft which equipped most of the Allied air forces, with more than 15 000 being built.

The RNZAF took delivery of just 30 P-51Ds in August – September 1945. Originally intended as the first of 370 to replace the Corsairs then being flown by the RNZAF's operational fighter squadrons, they proved to be the only Mustangs accepted. All contracts were cancelled when the war ended, and the aircraft were placed in storage.

In 1951 they emerged again, to be made airworthy for the Territorial Air Force squadrons being formed around the country, and for the next four years became a familiar sight around the four provincial squadrons' bases, particularly at weekends. Most of the pilots enjoying the Mustangs were, at first, World War II veterans, joined later by compulsory military training trainees who went on to regular aircrew courses, the majority of them airline pilots with NZNAC.

Ten Mustangs were written off in accidents and by 1955 the decision was made to retire all surviving aircraft — supposedly because of undercarriage defects, although that was widely regarded merely as an excuse to wind down the TAF (Territorial Air Force). Most of the remaining Mustangs were sold for scrap, although three went into private hands, two in a dismantled state and one (NZ2417) to be rejuvenated to fly once more in 1964 as ZK-CCG. For 10 years it was seen and heard in New Zealand skies before being exported to the USA, and this country was once more without the sound of a Merlin.

But many aviation people had the desire to see a Mustang active again. One of them was Tim Wallis, well-known for his helicopter exploits, who recalled watching these magnificent aircraft, operated from Wigram by No. 3 (Canterbury) Squadron TAF, during his boyhood years in Christchurch. Unable to prevent ZK-CCG leaving the country, he nevertheless kept a careful eye on the Mustang scene around the world as part of his wider vision of collecting a series of World War II inline engined fighters, and by the end of 1984 his plans were showing signs of success.

The Mustang he brought into New Zealand had been carefully inspected by Ray Mulqueen while under restoration, so the condition of every single part was known. Combining bits of ex-Indonesian Air Force (AURI) and ex-RCAF aircraft, it was shipped to New Zealand and reassembled at Wigram with expert help from RNZAF personnel who still remembered the type. Still under its US registration, the Mustang thrilled thousands of airshow spectators that summer and has continued to be one of the mainstays of the local warbird movement.

RNZAF Mustangs tended to be painted in unexciting colours, but ZK-TAF's paintwork is an exact replica of NZ2415, the aircraft flown by No. 3 Squadron's commanding officer, Squadron Leader Ray Archibald. For a short time the original NZ2415 sported a red propeller spinner, and was affectionately known as 'Rudolph the Red-nosed Reindeer' after a popular song of the day, until officialdom put a stop to this idiosyncrasy with an excuse that it might mesmerise groundcrews. It was withdrawn from service in January 1954 after losing power on approach to Wigram, so there's little chance that two identical P-51Ds will be seen in the sky at once.

ZK-TAF/NZ2415 went through a period of uncertainty recently. Changes in Tim Wallis's

The Warbird hangar at Ardmore can be crowded at times. Early morning sun greets the Mustang and many others.

Alpine Deer Group and some financial rearrangements, including the Spitfire project, saw the Mustang come up for sale. Unwilling to see it disappear back overseas, a group of concerned people led by Trevor Bland and Brian Rhodes formed the new Zealand Historic Aircraft Trust to buy the Mustang and, in the future, other famous aircraft and keep them flying in the country's skies.

With the aid of a bank loan and co-operation from Tim Wallis, the trust overcame legal delays and embarked on a major fund-raising campaign. Under its deed all money raised will be directed towards the purchase of historic aircraft and their maintenance and operation, keeping the Warbirds principle of airworthy and active aircraft very much to the fore.

BAGHDAD FURY

'Lightweight' is not a term normally associated with a single-seat aircraft weighing more than four tonnes empty, but that's how the Hawker Fury came into being. Designed by Sydney Camm as a 1942 response to the RAF's requirement for a lightweight version of the Tempest, it was modified the following year in line with the Royal Navy's specification for a fleet fighter.

With the war coming to an end, production of the land-based Fury was cancelled, as the jet fighter was clearly taking over from piston-engined aircraft. However, jets were still an unknown quantity for carrier operations, and the Sea Fury first flew in February 1945, powered by the biggest of Bristol's sleeve-valve family, the 18-cylinder 2 500 hp Centaurus, and entered naval service two years later.

The Sea Fury and its American contemporary, the Grumman F8F Bearcat, were the last of the piston-engined fighters. Both saw active service in Korea, and the Fury currently flying in New Zealand is painted in the colour scheme of one that shot down a Soviet MiG-15 jet fighter.

But the local warbird is not strictly a Sea Fury. An aggressive overseas sales campaign by Hawker resulted in Fury orders for Pakistan and Iraq, as well as Sea Furies for the Royal Australian Navy. Some 60 aircraft were eventually delivered to Iraq between 1950 and 1953, but little is known of their service life until their retirement in the early 1960s. Although retaining the wing-folding and tail-hook mechanisms of the shipborne fighters, they lacked the hydraulic operation, and became known as Baghdad Furies, similar in all other respects to those serving with various navies elsewhere. To complicate the issue, Iraq did receive some Fleet Air Arm Sea Furies diverted from the assembly line.

By the late 1970s some 27 Iraqi Furies had made their way to the USA, something of a coup in international warbird circles. Several years' hard work and diplomacy on the part of two Americans, some of it hindered by two Middle Eastern wars, saw a couple of shiploads of fuselages, wings, engines and miscellaneous spare parts rescued from the open desert conditions in which they had been standing for some years, and taken to Florida.

Four of them later went to Australia, joining the four ex-RAN Sea Furies still there after their retirement from active service and forming the mainstay of the Warbirds movement across

Opposite: New Zealand's only aircraft sporting a five-bladed propeller, the Fury is the last in a long proud line of British piston-engined fighters.

the Tasman. One arrived in New Zealand in 1986, to be completely stripped, refurbished and restored to flying condition, ready for its public debut at an airshow at Dairy Flat in March 1988.

ZK-SFR is the result of interest shown in a high-performance single-seat fighter by a consortium of Warbirds Association members who, unlike most other pilots of the faster aircraft in the fleet, have had no RNZAF or other experience of the heavier metal.

The New Zealand Baghdad Fury was made in 1952 and was one of the last delivered to Iraq, bearing the second-highest IAF serial number, 326. It is also one of the lowest in hours, just 120 from new, but naturally needed a full strip and rebuild after sitting for years in the desert's open storage. The 2 560 hp 18-cylinder sleeve-value Bristol Centaurus — still the original engine with 120 hours — was similarly stripped for inspection, but needed only some sleeves replaced, where they had been marked by being open to the atmosphere for many years, before careful reassembly.

Why a Fury? 'Because it was there,' says Rob Booth, one of the three flying members of the team. 'Well, not really — I'd seen them fly in the States and thought they were the most impressive aircraft. It was a good project, with parts available, and Bruce Coulter had already done one, working on the rebuild of G-FURY in England.' Bruce Coulter, chief engineer of Gulf Aeronautics in whose Ardmore hangar the Fury slowly took shape again, oversaw the project with help from numerous other organisations around Ardmore. Because nobody was qualified to fly the test schedule once it was ready for the air, Guido Zuccoli came across from Darwin, where his own ex-Iraqi Fury is based, and said he was most impressed with the standard of workmanship.

'Projects of this kind have been floundering in the United States,' he pointed out. 'It's an extremely competent team and the aeroplane is a real credit to them.' The Fury flying well, the test pilot returned home and the three flying members of the syndicate went to the United States to become current on a two-seater version. ZK-SFT's 1988 public displays at Dairy Flat and Wanaka were flown by Rob Booth, who had to cope with some teething troubles in the hydraulic system but shrugged off the brakeless landings as nothing to worry about. Instead, he described what it was like to fly a display routine in New Zealand's biggest and heaviest single-engined aircraft.

'I've had it up to 390 knots indicated, during a run at the Wanaka airshow,' he said. 'I would have gone faster, but the engine was still fresh and I didn't want to risk anything. It was fairly windy, too, and turbulent.

'The controls were still quite light at that speed. All controls are servo-assisted with tabs,

John Greenstreet in charge of plenty of Bristol horsepower.

and it's really a lovely aeroplane to fly. It's delightful, very manoeuvrable with no real nasty vices and very light in the controls. The stalls not vicious at all, and it tells you it's going to do it, with lots of buffet from the elevator and aileron trim tabs.'

It makes quite a change from the Harvards these pilots have been flying in Warbirds formation service, and there's nothing that will quite match the Baghdad Fury for sight and sound — not to mention fuel consumption, the big Centaurus engine burning 340 1/hr in normal cruise or 900 1/hr during an airshow routine. It may have been lightweight by 1940s Hawker standards, but it is a most impressive example of the swansong of piston-engined fighters.

Ownership of ZK-SFR has undergone some consolidation recently. Two of the original flying members of the group are pursuing their own warbird projects, and for a while it looked as though the Fury might have to be sold overseas, which would have been regretted after all the work that had gone into the project.

However, Rob Booth has now been joined by Grant Biel as major partner, while Mat Wakelin retains a minor non-flying share against the day when he, too, can take to the air in the Warbirds' biggest and most powerful fighter.

A famous silhouette in the early morning sun.

SPITFIRE

The Supermarine Spitfire holds a very special place in the hearts of many people. To the British it epitomises all that stood between them and defeat by the Germans during the Battle of Britain (which is not exactly fair on the workmanlike Hawker Hurricane, the Spitfire contemporary which actually bore the brunt of the fighting and went on to give a good account of itself in so many war theatres). The warbird enthusiast anywhere in the world will recognise the Spitfire for what it is — a superb design embodying the very best of aeronautical technology 50 years ago — and it has become an international legend in fighter aircraft history.

And the Spitfire rightly deserves its fame. Powered by numerous versions of the equally superb Rolls-Royce V-12 Merlin and the later and larger Griffon, it flew as well as it looked, and for a long time the ambition of every Allied fighter pilot was to fly a Spitfire. Many of them succeeded, too, for more than 20 000 Spitfires were made, as well as the navalised versions of Seafire and later Seafang and Spiteful.

The name Spitfire was originally given to an earlier Supermarine fighter project, a 1931 monoplane design with fixed undercarriage. By the time it entered service in small numbers, however, the company's chief designer, Reginald Mitchell, had started work on a new fighter. It was powered by Rolls-Royce's new PV-12 engine, developed from the Schneider Trophy-winning 'R' type, and Mitchell's design team also pooled its collective knowledge, from the Schneider Trophy racing days, to come up with a new fighter.

Its fuselage was given the smallest frontal area that would fit behind the V-12 engine, but what set it apart from its contemporaries and made it so distinctive in shape was the elliptical-plan wing. The necessity for a thin wing (to reduce drag for high performance) had to be offset against the RAF requirement to eight-gun armament, the four machine-guns and their ammunition in each wing taking up room in addition to the retractable undercarriage and engine cooling radiators. The answer was a broad-chord wing, tapering towards the tip to reduce drag, ideal aerodynamically but less so for mass production.

Nor was the Supermarine factory set up for making its new fighter in any quantity. Between 1919 and 1936 it had built a total of only 137 aircraft, most of them Walrus amphibians, and the new fighter was painfully slow to appear. The prototype first flew in March 1936, 15 months before Mitchell died of cancer at the early age of 42, but the first production example of what was known as the Spitfire did not appear until May 1938.

By the outbreak of war about one aircraft was being made per day, but the urgency of the situation saw production moved to other factories around Britain, with most Spitfires emerging from the Midlands facility at Castle Bromwich. Throughout the war the Spitfire was the only Allied fighter to be continuously produced, the last being made in 1947. During those 10 years the basic design saw increases of 100 per cent in engine power, 40 per cent in top speed, 80 per cent in rate of climb and 40 per cent in loaded weight. Such was the built-in development potential of the original aircraft that more than 40 major variants saw action, as well as a large number of minor versions for specific roles.

Although a large number of New Zealanders flew the Spitfire in combat, it was never in RNZAF service. Some Seafires from visiting aircraft carriers did fly in local skies during the 1940s, and a Spitfire has sat for many years in the Auckland War Memorial Museum, but the type is not a familiar sight in New Zealand and recent plans for a visiting example to take part in the airshow circuit did not come to fruition.

But Tim Wallis has changed all that. The man who brought an active Mustang to the New Zealand public in 1984 has imported his own Spitfire, so for the first time the distinctive shape has been seen — briefly — in some parts of the country.

ZK-XVI is a Mk.XVI with serial TB863, differing from the well-known Mk.IX only in having a Packard-manufactured Merlin in place of the Rolls-Royce engine. First seeing active duty in March 1945 with No. 453 Squadron RAAF, based at RAF Matlask, it flew with a number of squadrons, often on 'rent-an-enemy' anti-aircraft training roles, until July 1951 when the engine failed on take-off and the Spitfire was struck off charge and placed in storage.

Not long afterwards it started a career in movies, being used for cockpit shots in the 1955 film *Reach for the Sky*. Twelve years later it was bought for possible use in *Battle of Britain*. However, sufficient airworthy Spitfires meant that TB863, minus engine and much of its cockpit structure, was relegated to a static role, providing spares for its flying stablemates. It was later transferred to private ownership and restored to display state for the opening of the Historic Aircraft Museum at Southend in 1972.

Further changes of ownership followed, with work starting in 1982 on restoration to bring the Spitfire back to flying capability once again. By late 1985 it was at Duxford, part of Steven Grey's group now known as 'The Fighter Collection', and being worked on by David Lees whose previous experience included five Spitfire restorations.

The Spitfire's public debut took place at Ardmore in January 1989. Its wing makes a handy shelter from rain, but those people are soon going to have to run very quickly to keep up.

Grey's policy of variety within his collection led to the acquisition of a Bell P-63 Kingcobra and the release of Spitfire TB863, which was bought by Tim Wallis for his own collection of inline-engined fighters. It was completely finished and test flown at Duxford before being shipped out to New Zealand, with Marnix Pyle joining the restoration team to gain experience on the type in anticipation of its arrival and reassembly in this country.

Minor delays saw TB863/ZK-XVI make its public debut, in unpainted state, at an Ardmore airshow in January 1989, mere days after its first test flight in New Zealand. Steven Grey thrilled a large crowd with his immaculate demonstration of the Spitfire's handling, the low-level manoeuvrability very different from the Mustang's forte of high-speed passes and aerobatics taking up a large amount of sky.

Tim Wallis had been checked out in his new aeroplane and set out to fly it home to Wanaka, where it would be painted in preparation for its next display, at Wigram.

Unfortunately, things came to rest at Waipukurau. Setting up for a requested fly-past of the aerodrome, he suffered partial fuel starvation through running one tank low. He was unfamiliar with the Spitfire's complicated system and elected to make a forced landing on the airfield, but the engine stopped completely and TB863 ended up one paddock short, with damage to propeller and undercarriage.

The pilot was unhurt and the Spitfire salvaged and returned to Wanaka for major repairs. When next it appears it will be wearing full warpaint, and it will fill a very large gap in New Zealand aviation. ZK-XVI will be seen on appropriate occasions around the country for a long time, in recognition of the part played by this most famous of all piston-engined fighters.

One advantage of the older (and lighter) type of single-seat jet fighter is its ability to be moved around by hand.

DH112 VENOM

$\mathbf{T}$ he idea of a privately owned jet fighter, something capable of outperforming (in rate of climb, at least) the current equipment used by the country's air force, is something of a novelty to most people. Yet that describes Warbirds president Trevor Bland's de Havilland Venom, the first civilian jet fighter on the New Zealand civil register and the first jet operated by the Warbirds Association.

To the uninitiated the DH112 Venom looks much the same as the DH100 Vampire, familiar as the RNZAF's most prolific post-war aircraft, 58 of which were operated between 1951 and 1972 before they were replaced by the current Skyhawks and Strikemasters. The Vampire, de Havilland's single-seat interceptor, was first flown in 1943. Its distinctive twin-boom shape, accommodating the pilot in a small egg-shaped nacelle ahead of the DH Goblin turbojet engine, and the tailplane linking the twin rudders, still of de Havilland outline, were seen all over New Zealand.

Even before the Vampire first saw service with the RAF in 1946, a more powerful version was being planned using the DH Ghost, which promised 50 per cent more thrust from an engine only slightly bigger. Aerodynamic refinements were incorporated, mainly in the wing which was made thinner and given what was at the time termed 'sweepback', although only the leading edge was swept and the trailing edge remained straight.

The most obvious difference was the wingtip fuel tanks fitted as standard, and the new Venom also had wing fences to control airflow. The twin booms and de Havilland's wooden fuselage were retained, but the somewhat larger diameter engine resulted in a slight portliness in outline behind the cockpit. Its performance was shown by John Cunningham, the company's famous chief test pilot, who set an absolute world altitude record of 59 446 feet in March 1948, and after a number of defects of varying degrees of severity were cured, the DH112 Venom proved a successful fighter of the 1950s.

The RNZAF operated Venoms, but not in New Zealand. No. 14 Squadron flew the type after 1955 while based at Tengah in Singapore, but the aircraft were returned to the RAF when the squadron withdrew home in 1958.

Plans for the Venom to become the standard NATO fighter-bomber with assembly centres in France and Italy failed to eventuate because of the run-down of the British aircraft industry. One European country which did adopt the type, however, to the extent of manufacturing it under licence and flying it into the 1980s, was Switzerland. Some 250 Venoms were operated by the Swiss Air Force for more than 25 years, an achievement which few other jet fighters could rival.

Trevor Bland had flown Venoms in South-East Asia and had always harboured plans for a real warbird different from anything else. In 1985 he bought one of the Swiss aircraft, a high-level reconnaissance model which had been fully rebuilt by Pilatus in 1981 before being placed in reserve. In 1985 it was flown to Stansted in England, dismantled and shipped to New Zealand.

Unfortunately, reassembly was not as straightforward as expected, and the Venom took almost two years to be prepared for flight again. The major problem was damage during shipping, as the aircraft had not been secured properly inside its container; and some parts were missing.

But after a lot of work at Whenuapai by Gordon Phillips, assisted by Bill Rolfe, Peter Gardiner, Smoky Schrader, Graham Stanton and many others, the owner took his shiny Venom, painted to represent one of those operated in Singapore by No.14 Squadron, for its test flight in August 1987. Since then it has appeared in several airshows in the hands of John Denton, but frequency of flying is limited by the amount of fuel it consumes and the availability of starting cartridges.

Starting is spectacular to watch. An asbestos mat is placed over the tailplane to prevent paint scorching and the cordite cartridge gives a column of black smoke and a loud 'whoosh' as it turns the engine over. Fortunately it has proved reliable in that respect, to the relief of all concerned — especially at the big airshow at Wanaka in April 1988, when supplies were down to just one last cartridge. If that had failed to fire the engine, the Venom would have had to stay at Wanaka and John Denton hitch a ride home. The sighs of relief could be heard above the rising moan of the Ghost.

At this stage with an unsuccessful start, if the pilot hasn't remembered to close the canopy he's in for a bout of coughing. Pointing the Venom into wind can help.

REQUIEM

The warbird situation in New Zealand is never static. Quite apart from the actual aircraft ownership, which sees changes in the groups that operate and fly the individual warbirds, new types are occasionally imported and added to the growing fleet.

One gleaming example arrived at Ardmore towards the end of 1987 and was reassembled and flown a few days later. Alas, less than four weeks after the test flight it was a crumpled wreck in a Motueka orchard, too badly damaged to be rebuilt. That was caused by a mechanical problem, with the propeller changing to full coarse pitch just after take-off and not allowing the engine to develop enough power to stay airborne.

It marked the shortest flying career of any warbird in this country up to that time.

Tony Butcher imported his Cessna 195 from Vancouver, Canada. The model has the usual Cessna features of all-metal construction and high wing, but a 300 hp Jacobs radial engine on the nose sets it apart from the usual range of more modern types from the factory in Wichita, Kansas.

The Cessna 195 has a more rounded shape than its modern cousins.

Nearly 1200 Cessna 195s and the outwardly identical but less powerful 190s were made between 1947 and 1954. Some 83 earned true warbird status by being used by the US military forces as light cargo, VIP and ambulance aircraft in Alaska and South-East Asia.

The civilian examples have gained a reputation for being powerful, comfortable touring aircraft and are popular in the classic aircraft movement in the United States. None ever came to New Zealand when new, but several enthusiasts had expressed interest in the type and Tony Butcher was the first actually to achieve his aim.

And it's likely that more will follow — but not to emulate the example of ZK-MWY in every way. The aim of the New Zealand Warbirds Association is, after all, to keep its aeroplanes airworthy and airborne, where they belong.

APPENDIX

Harvard Mk.2A
Constructor's number: 88-10254
Year of manufacture: 1941
Current registration: ZK-ENN
Previous identity: NZ1025
Previous owners: RNZAF, D. Diamond,
   J. Sullivan
Type of operation: pilot training
Current owner: J. Greenstreet
Present base: Ardmore Aerodrome

Harvard Mk.3
Constructor's number: 88-14889
Year of manufacture: 1940
Current registration: ZK-ENF
Previous identity: NZ1065
Previous owner: RNZAF
Areas of operation: mostly Canterbury
Type of operation: pilot training
Current owners: syndicate of 15, four of whom
   flew it on their RNZAF wings course at Wigram
Present base: Ardmore Aerodrome

Harvard Mk.3
Constructor's number: 88-14672
Year of manufacture: 1942
Current registration: ZK-ENE
Previous identity: NZ1066
Previous owner: RNZAF
Type of operation: pilot training
Current owners: Rob Booth and Chris Porter
Present base: Ardmore Aerodrome

Harvard Mk.3
Constructor's number: 88-15873
Year of manufacture: 1942
Current registration: ZK-ENG
Previous identity: NZ1078
Previous owners: RNZAF, G. Martin,
    B. Dalliessi, J. Greville
Type of operation: pilot training
Current owners: group of 20
Present base: Ardmore Aerodrome

Dave Marshall

Harvard Mk.3
Constructor's number: 88-16505
Year of manufacture: 1943
Current registration: ZK-ENC
Previous identity: NZ1091
Previous owner: RNZAF,
Areas of operation: TAF Canterbury Squadron
Type of operation: pilot training
Current owner: Derek Williams
Present base: Tauranga Airport

Harvard Mk.3
Constructor's number: 88-16506
Year of manufacture: 1942
Current registration: ZK-WAR
Previous identity: NZ1092
Previous owners: RNZAF, Trevor Bland and
    Stan Smith
Type of operation: pilot training
Current owners: syndicate of 9
Present base: Ardmore Aerodrome

Dave Marshall

Harvard Mk.3
Constructor's number: 88-17004
Year of manufacture: 1942
Current registration: ZK-END
Previous identity: NZ1096
Previous owner: RNZAF
Areas of operation: Canterbury
Type of operation: pilot training
Current owners: Wally Bell, Rod Dahlberg,
Present base: Tauranga Airport

Harvard Mk.3
Constructor's number: 88-17010
Year of manufacture: 1942
Current registration: ZK-ENJ
Previous identity: NZ1098
Previous owner: RNZAF
Type of operation: pilot training
Current owners: group of 8
Present base: Ardmore Aerodrome

DH82A Tiger Moth
Constructor's number: DHNZ97
Year of manufacture: 1942
Current registration: ZK-AKC
Previous identity: NZ847
Previous owners: New Plymouth Aero Club
    (canopy Tiger Moth), T.G. Wells, J.V. Harvey,
    North Otago Gliding Club, J. R.G.Hanlon,
    T. Grant
Areas of operation: mostly Taranaki and North
    Otago
Type of operation: RNZAF, aero club, private,
    glider towing
Current owner: J.R. Crosbie
Present base: Ardmore Aerodrome

DH82A Tiger Moth
Constructor's number: 3795
Year of manufacture: 1938
Current registration: ZK-ALK
Previous identities: ZK-AGZ, NZ724
Previous owners: incl. I. East, M.J. Green,
    Auckland Gliding Club
Type of operation: aero club, RNZAF, private,
    glider towing
Current owner: J.R. Crosbie
Present base: Ardmore Aerodrome

DH82A Tiger Moth
Constructor's number: DHNZ91
Year of manufacture: 1941
Current registration: ZK-ALM
Previous identity: NZ841
Previous owners: incl. Tauranga Aero Club,
    Airspread, Airspray, Tauranga Gliding Club
Areas of operation: Tauranga, Masterton
Type of operation: RNZAF, agricultural, aero
    club, glider towing, private
Current owners: E.F. Doherty, D.W. Ives, J.L.
    and I.M. King, T.S. Newland
Present base: Ardmore Aerodrome

DH82A Tiger Moth
Constructor's number: 82906
Year of manufacture: 1936
Current registration: ZK-ANQ
Previous identities: R5011 (RAF), NZ892
Previous owners: incl. Kaitaia Aero Club,
    Adastra, D.A. Norman, R.L. Scrivener
Areas of operation: Northland, Bay of Plenty
Type of operation: RNZAF, aero club, private
Current owner: M. Burdan, R.A. Land,
    G.S. Thompson
Present base: Palmerston North Airport

DH82A Tiger Moth
Constructor's number: DHNZ22
Year of manufacture: 1940
Current registration: ZK-ASG
Previous identity: NZ772
Previous owners: R.E. Lawrence, Aircraft
    Services, N.E. Cunningham, D.J. Payne
Areas of operation: North Auckland, Manawatu
Type of operation: private, training
Current owners: W. Saunderson, B. Webb,
    W. West
Present base: Ardmore Aerodrome

DH82A Tiger Moth
Constructor's number: DHNZ88
Year of manufacture: 1941
Current registration: ZK-ATC
Previous identity: NZ838
Previous owners: incl. Kaitaia Aero Club,
    Northern Air Services, Wanganui Aero Club,
    Stratford Aero Club, J.D. Rees,
    J. C. Rasmussen, D. L. Sterling
Areas of operation: Northland, Taranaki, Waikato
Type of operation: RNZAF, aero club, private
Current owner: M.R.M.Broadbent (Queensland,
    Australia)
Present base: Ardmore Aerodrome (in storage)

DH82A Tiger Moth
Constructor's number: 82946 DHNZ15
Year of manufacture: 1940
Current registration: ZK-AUE
Previous identity: NZ765
Previous owners: Air Contracts, Aerodress
Areas of operation: Wairarapa, Hawkes Bay
Type of operation: RNZAF, topdressing
Current owner: B.R. Lynch
Present base: Dairy Flat Airfield

DH82A Tiger Moth
Constructor's number: 85349
Year of manufacture: 1941
Current registration: ZK-AUZ
Previous identities: DE315 (RAF), G-AJHR
Previous owners: Scottish Aviation, Wanganui
 Aero Work, Wanganui Aero Club,
 D.J. Billinghurst
Areas of operation: UK, Wanganui
Type of operation: RAF, agricultural, private
Current owner: Wanganui Aero Work
Present base: Wanganui Airport

DH82A Tiger Moth
Constructor's number: DHNZ128
Year of manufacture: 1942
Current registration: ZK-BLI
Previous identity: NZ1448
Previous owners: Hawera Aero Club, South
 Otago Aero Club, A.J. Padgett, D.J. McMillan,
 I.H. Atkinson
Areas of operation: Taranaki, Otago
Type of operation: RNZAF, aero club, private
Current owner: Sir Peter Elworthy
Present base: Craigmore, South Canterbury

DH82A Tiger Moth
Constructor's number: DHNZ139
Year of manufacture: 1943
Current registration: ZK-BRB
Previous identities: NZ1459, ZK-BRD
Previous owners: incl. New Plymouth Aero Club,
 I.H. Finlayson, R.G. Johnson
Areas of operation: Taranaki, Waikato, Auckland
Type of operation: RNZAF, aero club, spraying,
 private
Current owner: T. Wallis
Present base: Luggate Aerodrome

DH82A Tiger Moth
Constructor's number: DHNZ152
Year of manufacture: 1943
Current registration: ZK-BRM
Previous identity: NZ1472
Previous owners: R.J. & J.K. Harding, Valet
 Service (Nelson), K.H. Wells
Areas of operation: Wanganui, Nelson
Type of operation: RNZAF, private
Current owners: P. Beer, G. Bryham,
 L. Marshall
Present base: Ardmore Aerodrome

DH82A Tiger Moth
Constructor's number: 83393/DHNZ45
Year of manufacture: 1940
Current registration: ZK-BUO
Previous identities: NZ795, ZK-APS
Previous owners: incl. Southland Aero Club,
 South Canterbury Aero Club, I.M. Green, New
 Plymouth Aero Club, M. R. Falconer
Areas of operation: Southland, South
 Canterbury, Otago, Taranaki, Canterbury
Type of operation: RNZAF, aero club, private
Current owner: S.G. Spencer-Bower
Present base: Claxby, near Rangiora

DH82A Tiger Moth
Constructor's number: 84711
Year of manufacture: 1941
Current registration: ZK-CDU
Previous identities: T6296 (RAF), G-ALAD,
 ZK-BAW
Previous owners: RAF, Wiltshire School of
 Flying, Aerial Superspread, C.H.R. Liddell
Areas of operation: UK, Gisborne, Waikato,
 Auckland
Type of operation: RAF, agricultural, private
Current owners: P.E. Upton and W.W. Renouf
Present base: Ardmore Aerodrome

DH82A Tiger Moth
Constructor's number: DHNZ133
Year of manufacture: 1942
Current registration: ZK-CYC
Previous identities: NZ1453, INST168
Previous owner: M.P. Robertson
Areas of operation: Auckland
Type of operation: RNZAF, instructional
   airframe, private
Current owner: R.F. Duncan
Present base: Ardmore Aerodrome

Stearman PT-13D
Constructor's number: 75-5064
Year of manufacture: 16 August, 1943
Current registration: N4036
Previous identities: 42-16901, PI-C543, RP-C543
Previous owners: USAAF, PAF, etc.
Areas of operation: USA, Philippines, Hong
   Kong, New Zealand
Type of operation: military training, crop
   spraying
Current owners: Len and Wendy Cowper
Present base: Ardmore Aerodrome

J-3 CUB
Constructor's number: 2709
Year of manufacture: 5 January 1939
Current registration: ZK-AHC
Previous owners: various, including Nelson Aero
   Club, Ruapehu Aero Club and seven private
   owners
Areas of operation: throughout NZ
Type of operation: aero club, private
Current owners: Jim Peace, Auckland
Present base: Ardmore Aerodrome

L4B Grasshopper
Constructor's number: 9538
Year of manufacture: 1942
Current registration: ZK-ATU
Previous identities: 3677 (USAAF), VQ-FAG
   (Fiji)
Previous owners: USAAF, British Solomon
   Islands Government, L.W. Tarr &
   A.W. Brown
Areas of operation: Solomon Islands, Fiji,
   throughout NZ
Type of operation: military liaison/spotting, aerial
   spraying, aerial photography, private
Current owners: L.W. Tarr & B. Bell-Syer
Present base: Coroglen, near Whitianga

J/1N Aiglet
Constructor's number: 2157
Year of manufacture: 1946
Current registration: ZK-ATS
Previous identity: G-AIBV
Previous owners: incl. Nelson Aero Club,
   A.S. Ransby, D.B. Telford, P.G. McKellow
Areas of operation: Nelson, West Coast,
   Southland, Canterbury
Type of operation: aero club, private
Current owner: C. Glasgow
Present base: Ardmore Aerodrome

J/5 Adventurer
Constructor's number: 2892
Year of manufacture: 1949
Current registration: ZK-AXJ
Previous owners: incl. Stratford Aero Club,
   Taupo Aero Club, S. Pierce, C.N. Bellingham
Areas of operation: Taranaki, central North
   Island, Auckland
Type of operation: aero club, private
Current owners: group of 10
Present base: Ardmore Aerodrome

Isaacs Fury II
Year of manufacture: 1975
Builder: B. Thompson
Current registration: ZK-DMN
Previous owners: B. Thompson, O. Harnish
Areas of operation: Auckland
Current owner: C. Glasgow
Present base: Ardmore Aerodrome

Isaacs Fury II
Year of manufacture: 1982
Builder: J. Ross
Current registration: ZK-JHR
Previous owner: J. Ross
Areas of operation: Otago, Auckland
Current owners: G. Bartell, R. Carswell,
    P. Irvine, M. Jones, W. Reinauer,
    L. Woodgate
Present base: Hobsonville Aerodrome

Isaacs Fury II
Year of manufacture: 1979
Builder: R. Woods
Current registration: ZK-RFC
Previous owners: R. Woods, F. Renwick
Areas of operation: Canterbury, Auckland
Current owner: D. Simpson
Present base: Dairy Flat Airfield

Replica Plans SE5a
Year of manufacture: 1976
Builder: S. Tantrum
Current registration: ZK-SET
Previous owner: S. Tantrum
Areas of operation: Manwatu, Auckland
Current owners: Group of 11
Present base: Ardmore Aerodrome

Fokker Dr1
Year of manufacture: 1986
Builders: S. Tantrum, J. Lanham
Current registration: ZK-FOK
Previous owners: S. Tantrum, J. Lanham
Areas of operation: Manawatu, Auckland
Current owners: K. Skilling, C. Glasgow,
    J. Lanham, K. Trillo
Present base: Ardmore Aerodrome

DH89B Dominie
Constructor's number: 6662
Year of manufacture: 1943
Current registration: ZK-AKU
Previous identities: HG663 (RAF), NZ528
Previous owners: RAF, RNZAF, NZNAC,
    Nelson Aero Club, Rotorua Aero Club,
    Patchett Tours
Areas of operation: all over New Zealand
Type of operation: military communications,
    observer training and reconnaissance,
    scheduled airline service, tourist charter flights
Current owners: David Gray, Harry Norton
Present base: Dairy Flat Airfield

DH104 Devon 1B
Constructor's number: 04324
Year of manufacture: 1952
Current registration: ZK-KTT
Previous identity: NZ1808
Previous owners: RNZAF, D. Culham, B. Keay,
    J. Trewin
Areas of operation: Woodbourne, Wigram
Type of operation: navigational trainer
Current owners: group of 30
Present base: Ardmore Aerodrome

C-47B/DC.3
Constructor's number: 26480
Year of manufacture: 1943
Current registration: ZK-DAK
Previous identities: 43-49219 (USAAF), PI-
    C486, VH-PNM, VH-SBT
Previous owners: USAAF, Philippine Airlines,
    Patair, Bush Pilots Airways, Mackay Air
    Museum
Areas of operation: Far East, Papua New
    Guinea, Northern Territories/Queensland
Type of operation: military, airline
Current owners: Warbirds DC.3 Syndicate Inc.
Present base: Ardmore Aerodrome

DHC-1 Chipmunk
Constructor's number: C1/0064
Year of manufacture: 1950
Current registration: ZK-DUC
Previous identities: WB623 (RAF), VH-RVW
Previous owners: Royal Victorian Aero Club,
    Ken Scott
Areas of operation: England, Australia,
    Canterbury
Type of operation: flying training, aero club,
    private
Current owner: Simon Spencer-Bower
Present base: near Rangiora, Canterbury

DHC-1 Chipmunk
Constructor's number: C1/0018
Year of manufacture: 1950
Current registration: ZK-TNR
Previous identities: WB566 (RAF), G-AORR,
    HB-TUB
Previous owners: Cambridge University Air
    Squadron, No.4 Basic Flying Training School,
    Colton Aviation Services, R. Turner
Areas of operation: UK, Switzerland
Type of operation: flying training, private
Current owners: group of 8
Present base: Ardmore Aerodrome

DHC-1 Chipmunk
Constructor's number: C1/0834
Year of manufacture: 1952
Current registration: ZK-MUH
Previous identity: G-AMUH
Previous owners: Air Service Training, Royal Air
    Force College, D. Dallison
Areas of operation: UK
Type of operation: flying training
Current owner: J. Wall
Present base: Ardmore Aerodrome

de Havilland Canada DHC-2 Beaver
Constructor's number: 25
Year of manufacture: 1948
Current registration: ZK-CKH
Previous identities: ZS-DCG, G-ARTR
Previous owners: Tank Aircraft Pty Ltd, Air
    Services (Wairarapa), J.D. Menary
Areas of operation: Sierra Leone, Wairarapa,
    Northland
Type of operation: utility, topdressing
Current owners: group of 25
Present base: Ardmore Aerodrome

CT4B

Constructor's number: CT/4/B
Year of manufacture: 1972
Current registration: ZK-DGY
Previous owners: NZ Aerospace Industries Ltd,
    M.H. Nathan
Areas of operation: New Zealand to Europe
Type of operation: prototype test flying,
    demonstration, private
Current owners: group of 40
Present base: Ardmore Aerodrome

P-51D Mustang

Constructor's number: 122-41369
Year of manufacture: 1944
Current registration: ZK-TAF
Previous identities: 44-74829, RCAF9265,
    N8675E, N1769MD
Current owner: N.Z. Historic Aircraft Trust
Present base: Ardmore Aerodrome

Baghdad Fury

Constructor's number: 3723
Year of manufacture: 1952
Current registration: ZK-SFR
Previous identity: Iraqi Air Force 326
Previous owners: Iraqi Air Force,
    Ed Jurist and David Talichet.
Areas of operation: not known
Type of operation: military
Current owners: Rob Booth,Grant Biel, Mat
    Wakelin
Present base: Ardmore Aerodrome

Spitfire LF Mk. XVIe

Constructor's number: TB863
Year of manufacture: February 1945
Current registration: ZK-XVI
Previous identities: initially FU-P, also FB-Y, G-
    CDAN
Previous owners: RAAF, RAF, Metro-Goldwyn-
    Mayer, A.W. Francis, J. Parks, S. Grey
Areas of operation: Europe, UK
Type of operation: active wartime, anti-aircraft
    training
Current owner: T. Wallis
Present base: Luggate Aerodrome

DH112 Venom FB1

Constructor's number: 844
Year of manufacture: 1956
Current registration: ZK-VNM
Previous identity: Swiss Air Force J1634
Previous owner: Swiss Air Force
Areas of operation: Europe
Type of operation: high-altitude reconnaissance
Current owner: T.T. Bland
Present base: Ardmore Aerodrome